Tons of Trucks!

STREET SWEEPERS

By Luke Harasymiw

Gareth Stevens
PUBLISHING

Please visit our website, www.garethstevens.com. For a free color catalog of all our high-quality books, call toll free 1-800-542-2595 or fax 1-877-542-2596.

Cataloging-in-Publication Data

Names: Harasymiw, Luke.
Title: Street sweepers / Luke Harasymiw.
Description: New York : Gareth Stevens Publishing, 2020. | Series: Tons of trucks! | Includes a glossary and index.
Identifiers: ISBN 9781538245217 (pbk.) | ISBN 9781538245231 (library bound) | ISBN 9781538245224 (6 pack)
Subjects: LCSH: Street cleaning–Juvenile literature. | Trucks–Juvenile literature.
Classification: LCC TD860.H297 2020 | DDC 628.4'6–dc23

First Edition

Published in 2020 by
Gareth Stevens Publishing
111 East 14th Street, Suite 349
New York, NY 10003

Editor: Therese Shea
Designer: Katelyn E. Reynolds

Photo credits: Cover, p. 1 © iStockphoto.com/Willowpix; p. 5 Chystopoltseva Kateryna/Shutterstock.com; p. 7 DD Images/Shutterstock.com; pp. 9, 24 (brush) TFoxFoto/Shutterstock.com; pp. 11, 21 locrifa/Shutterstock.com; pp. 13, 24 (vacuum) Werner OTTO/ullstein bild via Getty Images; p. 15 Hadrian/Shutterstock.com; p. 17 Angela N Perryman/Shutterstock.com; p. 19 Elena Chevalier/Shutterstock.com; p. 23 brillenstimmer/Shutterstock.com.

Printed in the United States of America

CPSIA compliance information: Batch #CW20GS: For further information contact Gareth Stevens, New York, New York at 1-800-542-2595.

Contents

I love toy trucks.
I love real trucks!

This is a
street sweeper.
It cleans streets!

Some spray water.
Their brushes sweep.

These trucks take waste away.

This one has a vacuum!
It sucks up leaves.

126
WBO
Blust
WBO
WBO

Street sweepers
are loud!

BUCHER
CityFant 60
BUCHER

Some are small!

TENNANT
ATLV

Street sweepers clean in the morning. They clean at night. Some have lights!

RAVO

They clean up
after parades!

50

I love this truck.
It's useful!

Words to Know

brush

vacuum

Index